MATH CAN BE A DANGEROUS THING!

MATH CAN BE A DANGEROUS THING!

Harry Blair
Bob Knauff

Carolina™ Mathematics
Carolina Biological Supply Company
Burlington, North Carolina

MATH CAN BE A DANGEROUS THING!

© 1997 by Carolina Biological Supply Company
  2nd Printing, 2001

All rights reserved. No part of this book may be reproduced, stored in a retrieval system, or transmitted in any form or by any means, electronic, mechanical, photocopy, recording, or other means, without the prior written permission of the publisher. Teachers who own this book may copy cartoons for use with their classes.

Printed in the United States of America

ISBN: 0-89278-004-5

Carolina™ Mathematics
Carolina Biological Supply Company
2700 York Road
Burlington, NC 27215
USA
1-800-334-5551

MATH CAN BE A DANGEROUS THING!

N.A.S.T.* APPROVED

We all know that, in the wrong hands, math *can* be a dangerous thing resulting in answers that bear no relation to reality. Our second book of cartoons takes its name from that realization. *Math Can Be A Dangerous Thing!* casts another light-hearted look at some of the ways, both real and imagined, in which we encounter math in our world.

Our first book of cartoons, *Not Strictly by the Numbers*, was guaranteed to be at least $16\frac{2}{3}\%$ funny, and so far no one has seen fit to ask for their money back. Instead of a similar pledge for the humor of this new collection, we guarantee only that our cartoons are dietarily benign—they contain absolutely no calories, fat, or cholesterol.

Hopefully, these cartoons will bring you the healthful benefits of hearty laughs, chuckles, smiles, and perhaps even a groan or two, all of a mathematical sort. We hope to again diffuse some of the mystique of math and perhaps to make it less foreboding. Yes, math *can* be dangerous, but it can also be the source of fun and amusement as well as being a versatile tool for navigating life's problems. We trust our efforts will meet with your approval and prove to be dangerously amusing.

Harry Blair
Bob Knauff

* NATIONAL ASSOCIATION OF SKEWED THINKING

REAL ACKNOWLEDGEMENTS

Many thanks to Bev Benner, Kimm McCarthy, and Gray Amick for their skillful assistance in converting our loose cartoons into this book. Gray even suggested the ideas for three cartoons that made the cut.

Special thanks to our families and friends for their comments and suggestions upon being subjected to the cartoons during their development.

H.B. and B.K.

IMAGINARY ENDORSEMENTS

"I haven't laughed so hard since I put down Blair and Knauff's first book, *Not Strictly by the Numbers*."

Professor Cal Q. Lus
Math Department Head
Continental Divide College

"We've made *Math Can Be A Dangerous Thing!* required reading for all math majors."

Dr. Al Gebra
Dean
Integer State University

"Blair and Knauff have the wrong title, it should be *Math Can Be a <u>Hilarious</u> Thing!* I'd like to order 40 more copies for every math teacher in my district."

Ms. Fran Fraction
Math Curriculum Chairperson
Isosceles School District

100,000 BC
A SETBACK IN THE DEVELOPMENT OF THE WHEEL

AN UNFORTUNATE MIX-UP IN THE SIGN PAINTING DEPT.

"IT ALL BEGAN WHEN WE TAUGHT HIM TO BARK TO FOUR."

ALMOST IDENTICAL TWINS

THIS YEAR'S POLAR CO-ORDINATES

IF NEWTON HAD MISSED THE POINT...

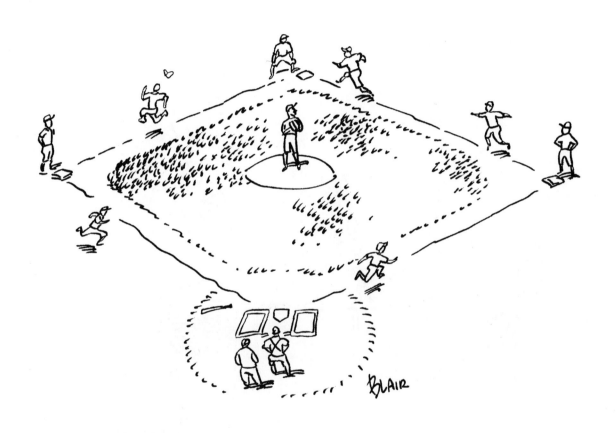

THE REDWINGS' RALLY GETS OUT OF CONTROL.

JENNIFER HAS FALLEN IN LOVE WITH A MATHEMATICIAN.

DICE FOR BEGINNERS

"BUT YESTERDAY 5 WAS 3 + 2!"

DESCARTES AT WORK

"THAT'S THE TROUBLE WITH A QUARTERBACK WHO'S A MATH MAJOR — INSTEAD OF A REVERSE, HE RAN A MULTIPLICATIVE INVERSE."

SHE WAS THE QUEEN OF COMPUTER NERDS.

ELVIS LIVES AS AN ELEMENTARY TEACHER IN TUPELO, MISSISSIPPI.

"RUBIK'S CONE"   "THE BERMUDA TRAPEZOID"

"ICE CREAM CUBE"   "GREAT CYLINDERS OF GIZA"

WHAT MIGHT HAVE BEEN...

PARTICIPANTS IN THE KNOT THEORY SYMPOSIUM

EARLY, SOMEWHAT FLAWED, BATTLE STRATEGY

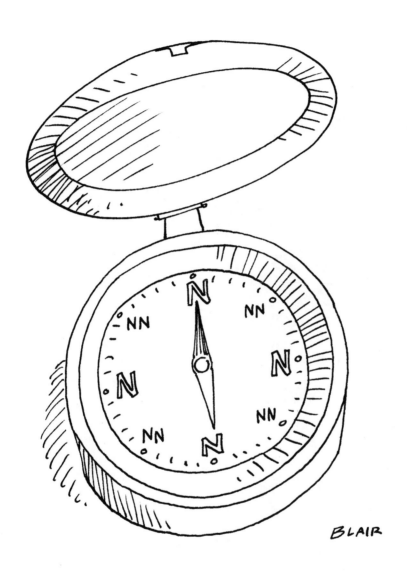

SOUTH POLE EXPLORER'S COMPASS

PANDORA'S BOX

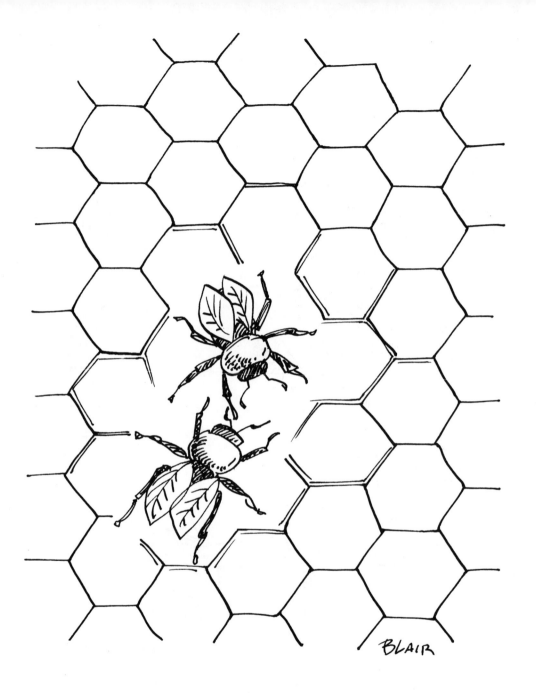

"THIS LOOKS FAMILIAR... THE BUSBYS MUST LIVE AROUND HERE SOMEWHERE."

"HEY! YOU'RE THE ONE WHO HAD TO HAVE 30-DAYS WRITTEN NOTICE TO BREAK THE LEASE!"

"TEACHER SAID I MISPLACED MY DECIMAL POINT."

"WE WOULD LIKE TO GIVE YOU THE ESTIMATED TIME OF ARRIVAL, BUT FRANKLY, NONE OF OUR PLANES ON THIS ROUTE HAVE EVER MADE IT."

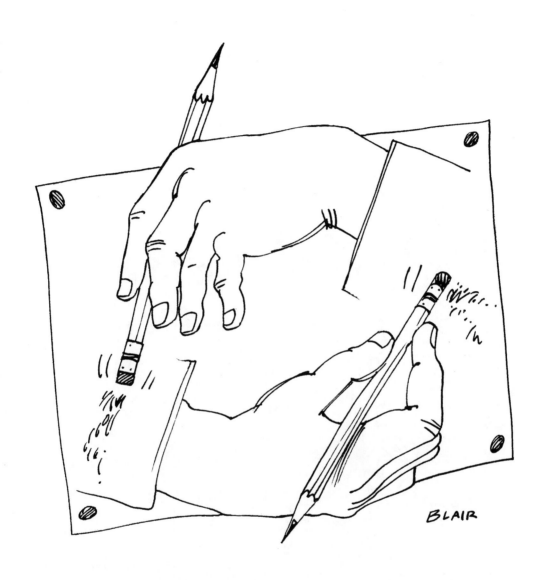

ERASING HANDS

THE RICE UNIVERSITY MATH DEPARTMENT'S TOUCH FOOTBALL TEAM

CAPTAIN JONES WAS A MATHEMATICIAN BEFORE HE WENT OFF TO SEA.

THE POLLSTERS AREN'T TAKING "NO" FOR AN ANSWER.

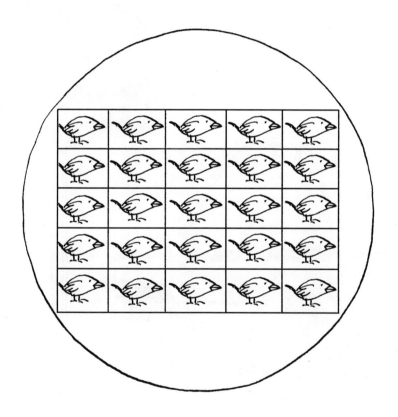

THE VIEW THROUGH 25X BINOCULARS

TO "HIT HIS WEIGHT," BUSTER BOGGS ELECTED TO LOSE WEIGHT RATHER THAN IMPROVE HIS SWING.

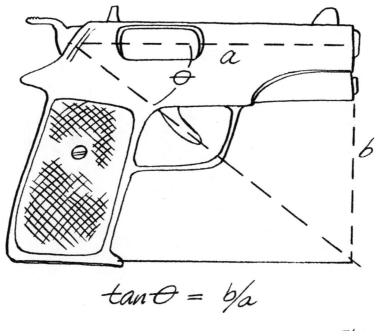

TRIGGER-NOMETRY

"I THINK A PASSING GRADE OF 100% IS GOING A BIT OVERBOARD ON THIS 'PERFECTION' STUFF."

NONTRADITIONAL SOLUTION

MATH DEPARTMENT RESTROOMS

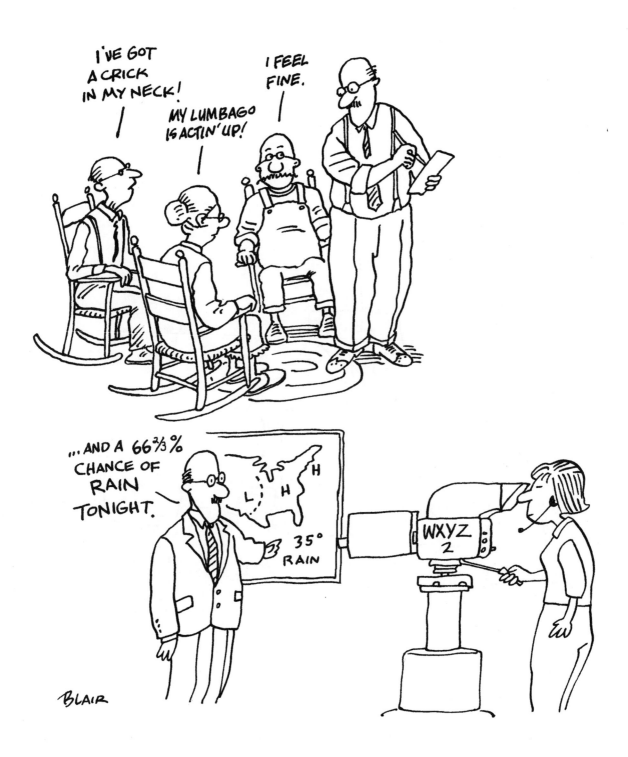

"YOU THINK YOU DESERVE PARTIAL CREDIT??!"

THE SPEED AND CONVENIENCE OF MASS TRANSIT IN THE ANIMAL KINGDOM

SO FAR, SO GOOD.

"'If straight line crosses two parallel lines, alternate interior angles will be equal!' What's yours say?"

GENERAL MATH, THE MILITARY MAN
REVERED BY MIDDLE SCHOOLERS EVERYWHERE

"POP" GIBSON WAS ALWAYS A CONTRARY CUSS.

HARRY'S CAPTION

"POP" GIBSON MARCHES TO A DIFFERENT DRUMMER.

BOB'S CAPTION

KNIGHTS OF THE MULTIPLICATION TABLE

"THE PAPER-GRADING PUNISHMENT WORKS ON EVERYONE BUT THE MATH TEACHERS."

"AND NOW, FOR HER TALENT, MISS MISSOURI WILL SOLVE $2x^2 - 7x + 3 = 0$ USING THE 'QUADRATIC FORMULA'."

"THE LINE AT CHECKOUT 7 IS SHORTER."

"IT'S THE SUPREME COURT, SAL. THE JUSTICES WANT THREE LARGE WITH EVERYTHING... AND THE MINORITY OPINION OF ONE SMALL WITH JUST PEPPERONI."

"WHEN DID YOU FIRST NOTICE THIS FEAR OF ODD NUMBERS?"

JULIE'S ALLOWANCE PIE CHART

ALL OPERATORS ARE BUSY. IF YOU WANT INFORMATION, PUSH "1"... IF YOU WISH TO PLACE AN ORDER, PUSH "2"... IF YOU HAVE A COMPLAINT, PUSH "873644239674."
BLAIR

"BUCK REALLY LIKES HIS NEW DOGHOUSE."

ALGEBRA ENTERTAINER

X, Y and Z Likely Tony Nominees
UNKNOWNS STAR IN NEW B'WAY HIT!

"Les Mathematiques" Plays to Packed House.

Story of 18th Century French Mathematicians Leads to Riots in Streets.

BLAIR

WHEN MATHEMATICIANS DINE OUT

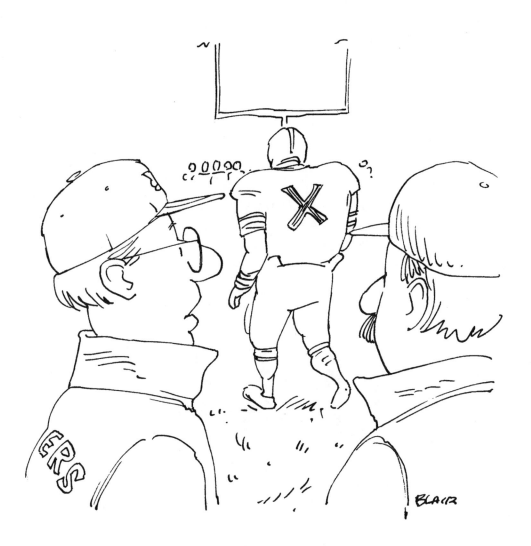

"THE TEAM AS A WHOLE IS REAL SOLID THIS YEAR... ONLY OUR NEW LINEBACKER IS AN UNKNOWN QUANTITY."

HERE'S THE SOLUTION TO THE PUZZLE ON THE PRECEDING PAGE.

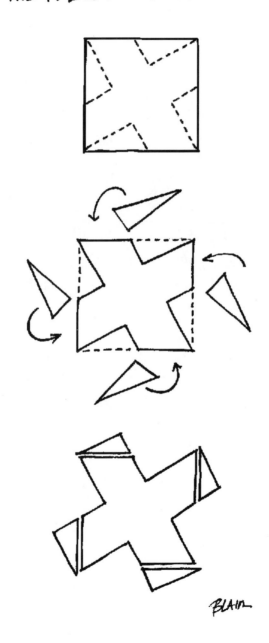

DISAGREEMENT OVER THE ORIGIN OF ALGEBRA

MATH GANGS

FLIRTING IN GEOMETRY CLASS

GNU MATH

"SURE IT WAS A NICE KISS, JIMMY, BUT DON'T TRY TO EXTRAPOLATE FROM THE LIMITED INFORMATION AT YOUR DISPOSAL."

"I HAVE THIS FUNNY FEELING THAT IT'S NOT GOING TO WORK."

IN ENGLAND, BUTCHER SHOPS ARE CONFUSING PLACES.

CLASSIC CASE OF TRYING TO PUT A SQUARE PEG IN A ROUND HOLE.

FIBONACCI LEARNS TO COUNT.

STONEHENGE DRUIDS TAKE SOLSTICE THAT NO ONE WAS HURT IN THE CONSTRUCTION ACCIDENT.

"THESE PROBLEMS MAY SEEM DIFFICULT AND BORING NOW, BUT JUST REMEMBER, YOU'RE LEARNING A SKILL YOU'LL PROBABLY NEVER USE AGAIN."

"THERE'S A 70% CHANCE OF RAIN TOMORROW. ... THAT'S 21% CELSIUS."

THROWING OUT THE HIGHEST AND LOWEST SCORES, DAD RECEIVES AN 8.8 FOR THE EVENING MEAL!

"TRICIA ALWAYS GETS HIGH MARKS FOR CREATIVITY."